Weird and Wonderful Animals

BLOWFISH

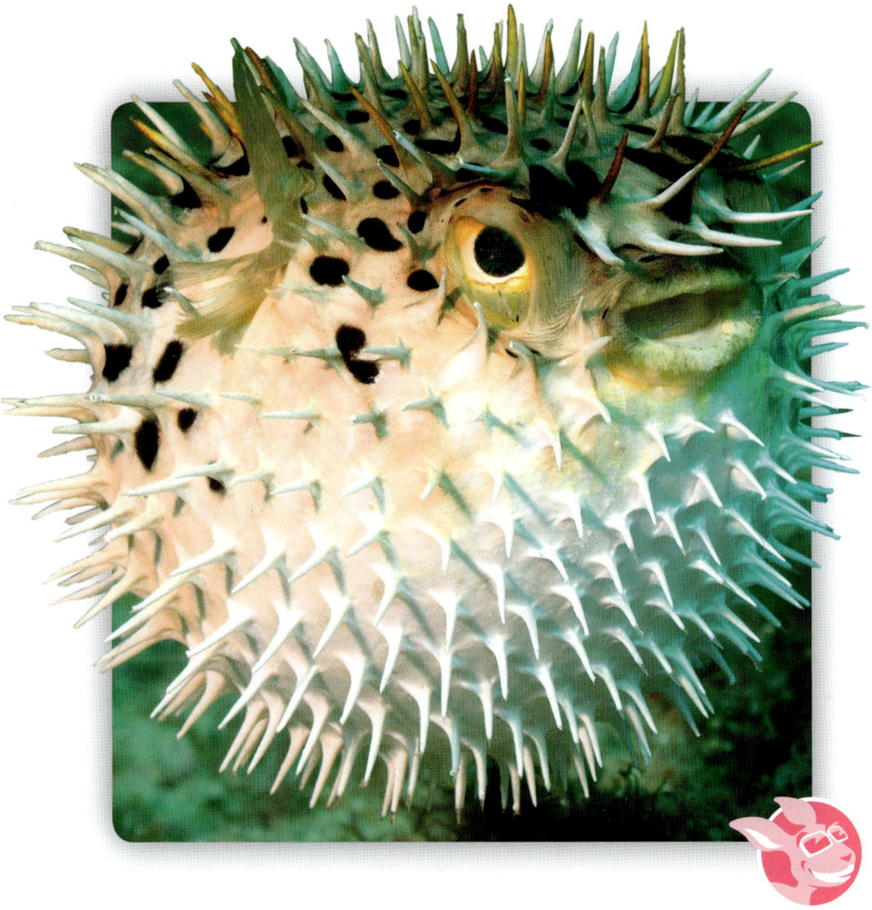

Emma Bassier

DiscoverRoo
An Imprint of Pop!
popbooksonline.com

abdobooks.com

Published by Pop!, a division of ABDO, PO Box 398166, Minneapolis, Minnesota 55439. Copyright © 2020 by POP, LLC. International copyrights reserved in all countries. No part of this book may be reproduced in any form without written permission from the publisher. Pop!™ is a trademark and logo of POP, LLC.

Printed in the United States of America, North Mankato, Minnesota.

102019
012020

THIS BOOK CONTAINS RECYCLED MATERIALS

Cover Photo: Shutterstock Images
Interior Photos: Shutterstock Images, 1, 6, 14, 15, 16 (top), 17 (bottom), 21, 23; iStockphoto, 5, 8, 11, 13, 16 (bottom), 17 (top), 19, 20, 22, 24–25, 30, 31; Fred Bavendam/Minden Pictures/Newscom, 7; Red Line Editorial, 9; Norbert Probst imageBroker/Newscom, 12; Georgette Douwma/Science Source, 27; Michael Patrick O'Neill/NHPA/Photoshot/Newscom, 28; Paulo de Oliveira/NHPA/Avalon/Newscom, 29

Editor: Nick Rebman
Series Designer: Jake Slavik
Library of Congress Control Number: 2019942490
Publisher's Cataloging-in-Publication Data

Names: Bassier, Emma, author.

Title: Blowfish / by Emma Bassier

Description: Minneapolis, Minnesota : Pop!, 2020 | Series: Weird and wonderful animals | Includes online resources and index.

Identifiers: ISBN 9781532166044 (lib. bdg.) | ISBN 9781644943342 (pbk.) | ISBN 9781532167362 (ebook)

Subjects: LCSH: Blowfishes--Juvenile literature. | Puffers (Fish)--Juvenile literature. | Oddities--Juvenile literature. | Fishes--Behavior--Juvenile literature. | Marine animals--Juvenile literature.

Classification: DDC 597.64--dc23

WELCOME TO
DiscoverRoo!

Pop open this book and you'll find QR codes loaded

with information, so you can learn even more!

Scan this code* and others

like it while you read, or visit

the website below to make

this book pop!

popbooksonline.com/blowfish

*Scanning QR codes requires a web-enabled smart device with a QR code reader app and a camera.

TABLE OF CONTENTS

CHAPTER 1
UNEXPECTED ABILITY

A blowfish swims in shallow water.

A larger fish swims toward it with an

open mouth. The blowfish puffs up its

body. Suddenly, the blowfish looks much

WATCH A VIDEO HERE!

When a blowfish puffs up, other fish may decide it is too big to eat.

bigger. **Spines** stick up all over its body.

The large fish swims away. The blowfish

is safe.

There are more than 20 kinds of porcupinefish.

Blowfish are also known as pufferfish. Scientists have discovered more than 200 types of blowfish. Some types have stiff spines. They are called porcupinefish. Other types do not have spines. But they can all puff up to defend themselves from **predators**.

Most types of blowfish do not have spines.

Blowfish live all around the world.

Many live in **tropical** water near coasts.

DID YOU KNOW?

When a blowfish puffs up, it can be up to three times its normal size.

RANGE MAP

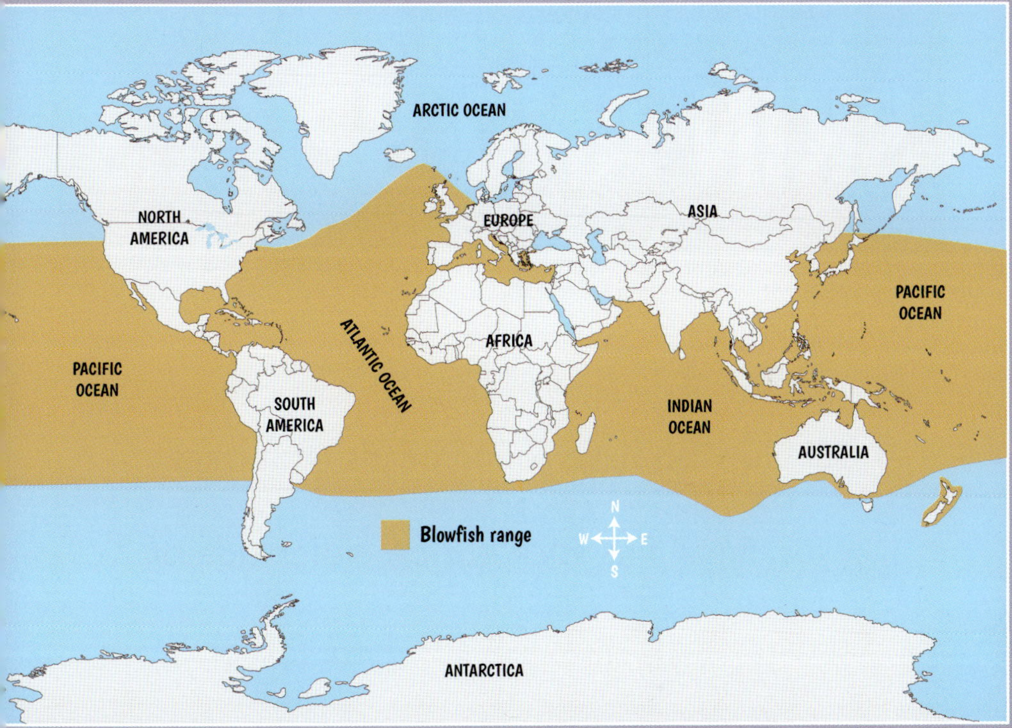
ARCTIC OCEAN
NORTH AMERICA
EUROPE
ASIA
PACIFIC OCEAN
ATLANTIC OCEAN
AFRICA
PACIFIC OCEAN
SOUTH AMERICA
INDIAN OCEAN
AUSTRALIA
Blowfish range
N
W
E
S
ANTARCTICA

The water in those locations is warm and shallow. Blowfish often find shelter in rocks.

CHAPTER 2
SKIN AND SPINES

All blowfish have round heads with large eyes. Small fins stick out on either side of their bodies. Blowfish also have fins near their tails. These fins are on the top and bottom of their bodies.

LEARN MORE HERE!

A blowfish swims past a coral reef.

A diver sees a large pufferfish.

Blowfish range in size. The smallest are only 1 inch (2.5 cm) long. But large blowfish can grow up to 3 feet (0.9 m) in length.

Unlike many fish, blowfish do not have scales. Instead, their bodies are covered in rough skin or **spines**. The skin on their bellies is stretchy. This allows them to inflate.

A porcupinefish's spikes lay flat when its body is not puffed up.

A blowfish shows off its hard beak.

A blowfish has a hard beak. The

beak is made of two or three teeth that

are **fused** together. Like other fish, a blowfish breathes through gills.

DID YOU KNOW?

A blowfish's beak never stops growing. But frequent biting keeps the beak from getting too long.

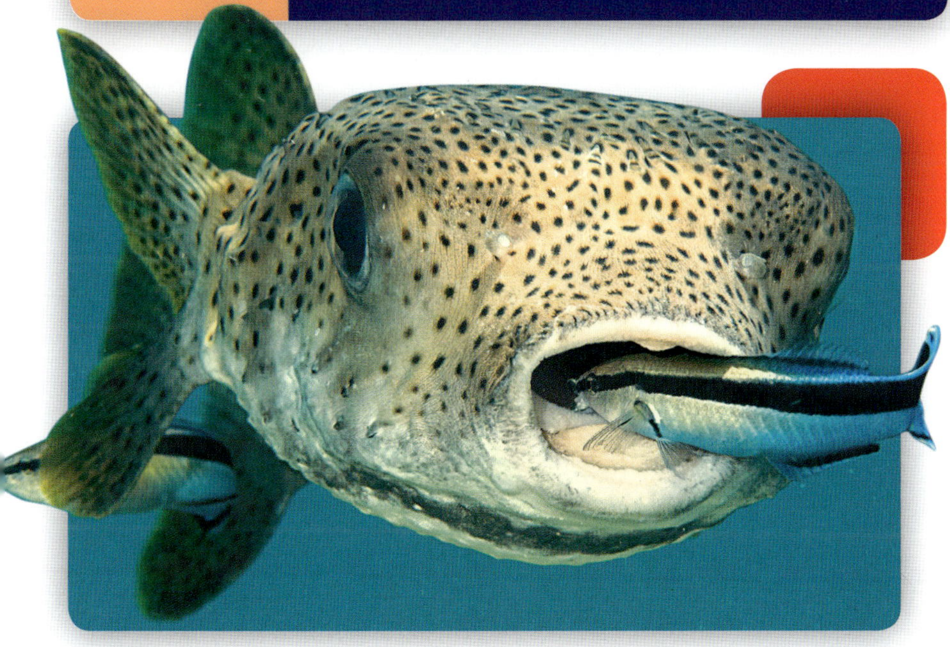

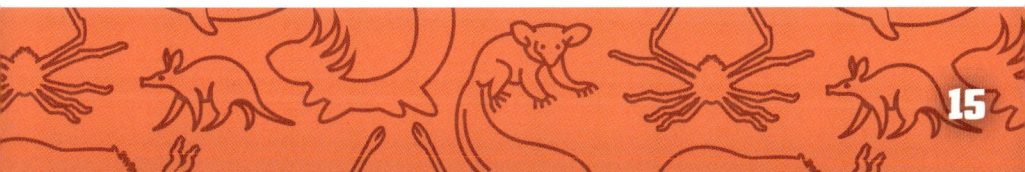

Some blowfish allow small fish to swim inside their mouths to clean them.

LIFE CYCLE OF A BLOWFISH

A female blowfish lays eggs in fine sand.

Most blowfish live for four to ten years.

The male fertilizes the pile of eggs after the female leaves.

The eggs hatch.

The young blowfish grow larger and become adults.

CHAPTER 3
PUFFING UP

Animals such as sharks, otters, and eels

eat blowfish. A blowfish uses its large

eyes to spot **predators**. Sometimes

it has time to hide in the cracks of

LEARN MORE HERE!

A blowfish puffs up when it senses danger.

rocks. But most blowfish are not fast

swimmers. Instead, they puff up their

bodies as defense.

A blowfish quickly sucks in water. Its belly fills, and the stretchy skin expands. Puffing up makes the fish look dangerous.

Some predators may be frightened or confused when a blowfish puffs up. They may swim away instead of attacking. Or, if a predator is biting the blowfish, it might let go.

A large fish attempts to eat a porcupinefish.

Some people eat blowfish after carefully removing the toxin.

In addition, most blowfish have a **toxin** in their bodies. This substance makes blowfish taste bad to predators.

The toxin can cause a predator to stop breathing and become **paralyzed**. In many cases, the toxin is deadly.

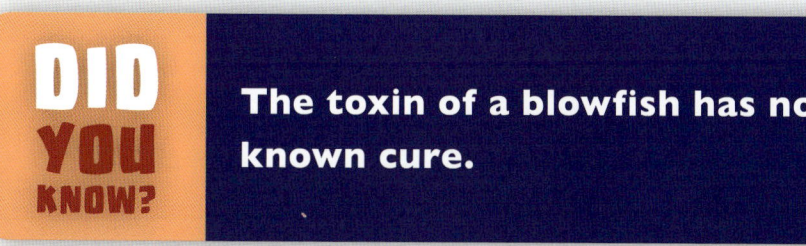

Bright colors warn predators that this blowfish is poisonous.

Puffing up takes a lot of energy.

Blowfish are tired for a few hours after

deflating. They are at greater risk

of being eaten during this time. If a

A blowfish hides near rocks to avoid predators.

predator attacks, the blowfish may not

have enough energy to escape. Blowfish

often hide to stay safe after deflating.

CHAPTER 4
SURVIVING

A blowfish eats **bacteria**, sea stars, and shellfish. The blowfish uses its hard beak to crack open shells. Many blowfish live in groups. A group of blowfish is called a school.

COMPLETE AN ACTIVITY HERE!

A blowfish snacks on a sea star.

DID YOU KNOW?

Piranhas are known for biting with razor-sharp teeth. Some blowfish bites are just as dangerous.

A blowfish blends in with the seafloor.

Some blowfish have dull skin colors. These fish blend in with their environment. Blending in helps the blowfish hide from **predators**. Other blowfish have bright skin colors. These colors warn other animals that the blowfish are poisonous.

MATING PATTERNS

Some blowfish create circular patterns in the sand to attract mates. Small male fish can create patterns that are 6.5 feet (2.0 m) wide. The pattern takes a week to make. Water flowing over the pattern gathers fine sand in the middle. This impresses the female fish. Females need fine sand to lay eggs in.

A blowfish creates a pattern in the sand.

MAKING CONNECTIONS

TEXT-TO-SELF

Have you ever seen a fish swimming in the wild? If yes, where were you? If not, where could you go to see fish?

TEXT-TO-TEXT

Have you read books about other animals that live in the ocean? How are those animals similar to or different from blowfish?

TEXT-TO-WORLD

Blowfish defend themselves by puffing up. How do other animals defend themselves?

GLOSSARY

bacteria – tiny life-forms.

deflate – to let out air or water and get smaller in size.

fuse – to grow together into one.

paralyzed – unable to move.

predator – an animal that hunts other animals for food.

spine – a sharp point that grows out of an animal's skin.

toxin – a type of poison made by living things.

tropical – having to do with a place where the weather is usually warm and wet.

INDEX

ONLINE RESOURCES
popbooksonline.com

Scan this code* and others like it while you read, or visit the website below to make this book pop!

popbooksonline.com/blowfish

*Scanning QR codes requires a web-enabled smart device with a QR code reader app and a camera.